Sensors 2017

Topical Conference at the 2017 AIChE Annual Meeting

Minneapolis, Minnesota, USA
29 October – 3 November 2017

ISBN: 978-1-5108-5787-2

Printed from e-media with permission by:

Curran Associates, Inc.
57 Morehouse Lane
Red Hook, NY 12571

Some format issues inherent in the e-media version may also appear in this print version.

Printed by Curran Associates, Inc. (2018)

For permission requests, please contact AIChE
at the address below.

AIChE
120 Wall Street, FL 23
New York, NY 10005-4020

Phone: (800) 242-4363
Fax: (203) 775-5177

www.aiche.org

Additional copies of this publication are available from:

Curran Associates, Inc.
57 Morehouse Lane
Red Hook, NY 12571 USA
Phone: 845-758-0400
Fax: 845-758-2633
Email: curran@proceedings.com
Web: www.proceedings.com

TABLE OF CONTENTS

Some abstracts in this book refer to images that were not made available by the publisher

506505 The Development of a Sensitive Electrochemical Method for Carotenoid Detection

Abstract:

Carotenoids are biological pigment molecules that are often found in fruits, vegetables, and human tissue. Humans cannot synthesize carotenoids, but they can be accumulated through diet (Fiedor & Burda, 2014). Carotenoids have been found to lower oxidative stress which can lead to a lower risk of disease and infections (Mares-perlman et al., 2002). For example, lower carotenoid levels have also been found in patients with Alzheimer's Disease, Parkinson's Disease, and multiple types of cancer (de Munter et al., 2015; Nolan et al., 2014). Although there is a reverse correlation between the concentration of carotenoids and risk of disease, there are still a lot of qualities of carotenoids that are unknown to scientists. Current methods to evaluate carotenoids in serum include Raman spectroscopy, high pressure liquid chromatography, and mass spectroscopy; however, all of these methods require large equipment, and therefore, not point-of-care.

We aim to develop an effective point-of-care, cheap, and sensitive carotenoid sensor; electrochemical methods are cost effective and can maintain sensitivity. Two electrochemical methods, cyclic voltammetry and electrochemical impedance spectroscopy, were used for the detection of beta-carotene and lutein (two carotenoids). Using cyclic voltammetry, we were able to calibrate our glassy carbon electrode sensor on the micro-molar scale. Electrochemical impedance spectroscopy, which measures the surface impedance, can be a more sensitive technique if the analyte adsorbs to the surface. We successfully detected nano-molar levels using electrochemical impedance spectroscopy, increasing sensitivity 100 fold. In addition to talking about our abilities to detect carotenoids using these two techniques, we will also discuss some of the reproducibility issues and control experiments conducted.

References:

Fiedor, J., & Burda, K. (2014). Potential Role of Carotenoids as Antioxidants in Human Health and Disease. Nutrients *6*, 466–488.

Mares-perlman, J.A., Millen, A.E., Ficek, T.L., & Hankinson, S.E. (2002). Symposium : Can Lutein Protect Against Chronic Disease ? The Body of Evidence to Support a Protective Role for Lutein and Zeaxanthin in Delaying Chronic Disease . Overview 1 , 2. 518–524.

de Munter, L., Maasland, D.H., van den Brandt, P.A., Kremer, B., & Schouten, L.J. (2015). Vitamin and carotenoid intake and risk of head-neck cancer subtypes in the Netherlands Cohort Study. Am. J. Clin. Nutr. *102*, 420–432.

Nolan, J.M., Loskutova, E., Howard, A.N., Moran, R., Mulcahy, R., Stack, J., Bolger, M., Dennison, J., Akuffo, K.O., Owens, N., Thurnham, D.I., & Beatty, S. (2014). Macular Pigment, Visual Function, and Macular Disease among Subjects with Alzheimer's Disease: An Exploratory Study. J. Alzheimer's Dis. *42*,

1191–1202.

501377 A Radio-Colorimetric Hydrogel for Detection of Therapeutic Levels of Ionizing Radiation Using Plasmonic Nanoparticles in 3D

Abstract:

Modern radiotherapy uses complicated techniques to deliver radiation dose to the tumor in hopes of sparing the healthy tissue. Improvements in radiation techniques has led to minimization of the side effects in cancerous patients. To improve patient wellbeing and quality it is important to monitor and record the dose delivered. Existing sensors are complex and require extensively trained professionals to operate them. In this talk, we discuss a novel colorimetric sensor that can detect therapeutic levels of ionizing radiation. We accomplished this by using ionizing radiation along with cationic micelles, in order to convert colorless Au^{+} ions to a pink colored dispersion of gold nanoparticles incorporated in a hydrogel matrix. Upon increasing the radiation dose, a significant visual increase in intensity of the developed color was observed. The difference in color intensities developed was used as a qualitative indicator of radiation dose and Ultraviolet-Visible spectroscopy was used a quantitative measure. In addition, a fixing agent is allowed to diffuse into the hydrogel and render spatial dose monitoring capability to the nanosensor. The plasmonic nanosensor is able to detect radiation doses in the therapeutic range, making it clinically viable. Our results taken together show a strong potential to be used as an invitro/invivo sensor for verification of dose delivered during radiotherapy in a clinical setting.

498156 Wireless Measurement of Enzymatic Degradation Kinetics with a Resonant Antenna Biosensor

Abstract:

This work focuses on the development of a chipless radio frequency biosensor that can be embedded both in or on materials for wireless monitoring of biologic activity. The coil antennas are patterned on copper coated polyimide substrates and etched with a solution containing hydrogen peroxide and hydrochloric acid. The antennaâ€™s initial resonant frequency is designed by the coil length and spiralâ€™s pitch and is tuned for the 1-100MHz range (for better penetration through soil, water, and tissue).Â The resonant frequency is measured at a distance by a two-loop coil attached to a vector network analyzer monitoring the S21 parameter. The resonant frequency shifts with any change in capacitance between the antenna coil caused by variation in dielectric permittivity of the medium on top of the antenna. In this research effort, we apply this sensor platform to study the activity of protease degradation of gelatin materials. The wireless sensors allow us to measure the effect of gelatin content, plasticizer weight percentage, and enzyme concentration on the overall degradation profile of the gelatin film in a simple, multiplexed experiment. We correlate these results to the mechanical properties of the gelatin films. We also discuss extensions of these sensors in other applications where measuring biologic activity in contexts where lack of optical and electrical connection impede existing tools, such as monitoring the contents of a disposable bioreactor.

498410 Thermodynamic Control of Response in Ionophore-Based Optical Nanosensors

Abstract:

Optical nanosensors are used to detect a wide range of ions and molecules by changing their fluorescent properties in response to the local analyte concentration. Optical nanosensors have key attributes such as biocompatibility and high spatial and temporal resolution that will allow them to be put toward clinical use and answer scientific questions in a way not possible with conventional electrochemical sensors. However, optical nanosensors have narrower dynamic ranges than electrochemical sensors, and thus practical methods to adjust the sensor response characteristics are needed to match the sensor linear range with the expected analyte fluctuation for an application. Adjusting the dynamic range includes shifting the upper and lower limits and expanding the span of the range. In ionophore-based optical sensors (used to detect ions), the dynamic range is determined by three simultaneous equilibrium expressions: extraction of the analyte from aqueous media into the organic core of the sensor, analyte binding to an ionophore, and acid dissociation of a pH indicator. In this work, we determine the effects of adding a fourth equilibrium by including either a second pH indicator or second ionophore to the typical formulation. The conventional ionophore-based optical sensor response equation is expanded to include a fourth equilibria and used to predict dynamic range control and extension at different sensor compositions. We show that the thermodynamics of chromoionophore acid-dissociation and analyte/ionophore binding are fundamentally different when two pH indicators are loaded within the same sensor matrix, effectively coupling their equilibria, as opposed to being housed in separate sensor nanoparticles. We demonstrate that these two methods of dual-component sensor design give finer control over the response range of ionophore-based optical sensors with than previous methods and can extend the linear range span up to 47% over what is possible with a single-chromoionophore nanosensor. This work shows, as a proof of concept, novel tools for range control and extension that can be applied toward designing an optical ion sensor with a dynamic range matching electrochemical sensors.

500375 Combining Forward Osmosis with Electrochemistry to Detect Ultra-Low Concentrations of Bacterial Virulence Factors and Quorum Sensing Molecules in Bodily Fluids

Abstract:

Current diagnostic methods for bacterial infection require the use of culture plates, a process that takes 1 to 5 days before clinicians can confirm a bacterial infection and provide the required antibiotic to the patient. This delay in acquiring results can hinder patient care and contribute to the growing impact of antibiotic resistant bacteria. Therefore, rapid detection of bacterial infections at point-of-care would play a key role in improving patient care and potentially prevent bacterial antibiotic resistance. In this work, we present how electrochemistry can be coupled with low cost forward osmosis to detect ultra-low concentrations of unique bacterial quorum sensing molecules in bodily fluids to serve as an early stage indicator of the identity of the potential bacterial infection. In this case, the clinician can provide targeted antibiotics instead of waiting for a culture plate to determine the type of bacterial infection.

Disposable carbon-based electrochemical sensors were used to run square-wave voltammetry on liquid samples of bacteria. *Pseudomonas aeruginosa* was used as the model bacterial species due to its production of a unique redox active molecule, pyocyanin that can be detected electrochemically. Differing bodily fluids were spiked with known concentrations of pyocyanin that were initially undetectable with the bare carbon sensors. The sample was then concentrated using a low-cost forward osmosis setup and then electrochemically rescanned to determine the final detectable pyocyanin concentrations. The advantage of this setup is its low cost, reproducibility, and accuracy as compared to current sample concentration methods that involve centrifuges and/or evaporators. A calibration curve was performed to determine the time-dependent concentration of sample in the forward osmosis setup and was used to calculate the initial starting concentrations of pyocyanin in the sample prior to using forward osmosis. The results indicate the feasibility of utilizing forward osmosis alongside electrochemistry towards detecting low levels of bacterial virulence factors in bodily fluids.

499145 Healthcare Technology Platforms: Engineering a Cellulose Strip for Biomarker Detection

Abstract:

Cellulose based microfluidic devices have shown great promise as a means of affordable and reliable disease diagnostics in developing countries. Despite an increase in the published literature, innovative approaches in fabrication and design of paper fluidic devices the knowledge to address the intricacies of device design for simple to complex reaction schemes remains at a nascent stage. Such designs require a rational approach based on the precise engineering of reactions and transport phenomena which are essential to ensure the optimum performance of the biosensor. Here, we have integrated and demonstrated different rational design frameworks and in particular, we considered the effect of hydraulic resistances and accordingly tuned the device dimensions to achieve the flow rates of specified reagents. In the process, we experimentally verified remobilization behavior and modelled the flow profiles for dry reagents stored on the paper fluidic chip. By taking the available kinetic models of antibody and antigen interactions further studies were carried to see the effect of rational design approaches on reaction and transport of analytes and the collective outcome of the colorimetric signal. As on alternative procedure to either conventional standalone or cellphone based image processing techniques which may suffer due to ambient conditions, a simple RGB color sensor capable of sensing even weak signal without interference of stray light was developed. This sensor is inexpensive, robust, and provide immediate results, thereby enabling quantitation without the need of expensive image processing techniques/devices. The complete technology platform which includes test strip engineering and development of RGB sensor was used for demonstrating the colorimetric detection of Osteocalcin, a biomarker for Osteoporosis. Efforts are on way to make this paper microfluidic technology platform to be generic and used for detecting panel of biomarkers making a way for low cost and efficient sensing technologies for disease diagnosis and prognosis.

497361 A Novel Modified DOT Blot Approach for Early Detection of Osteoporosis on Cellulose Substrates

Abstract:

As the world population is aging at faster rate there is a growing number of lifestyle and non-communicable diseases severely affecting the quality of life in elderly people. Therefore, there is a great need of developing novel technological solutions to address these challenges for improving the quality of life. Among many, one such condition which is severely afflicting the lives of one in three women and one in five men aged fifty years and above is Osteoporosis, a silent killer. According to a survey, it is responsible for causing over 8.9 million fractures worldwide and despite its debilitating nature, the tests to detect Osteoporosis and therapeutic monitoring of the disease is not performed routinely due to high costs involved. The conventional methods use dual energy X-ray absorptiometry (DXA) usually located in centralized laboratories to measure the bone mineral density (BMD) costing up to 250 dollars per test which is unaffordable by many and hence not suitable for point of care testing and therapeutic monitoring of the disease. Therefore, it is imperative to develop a novel alternative method for early detection and monitoring of Osteoporosis. Here, we have developed an immunological technique by modifying the DOT BLOT method to estimate the bone turnover markers (i.e. bone absorption and resorption markers) such as Osteocalcin present in the blood on cellulose based substrates. The method consists of a colorimetric based indirect enzyme linked immunosorbent assay (ELISA) capable of detecting the Osteocalcin down to 6 ng/ml. The novel procedures being implemented here to modify the DOT BLOT method required extremely low reagent volumes and detects the protein markers in less than two hours of time as opposed to eight hoursâ€™ conventional procedure. Most importantly the newly developed procedure for detecting Osteoporosis on cellulose substrate with shorter analysis times to carry out at point of care settings significantly reduces the cost per unit test which further allows clinicians to monitor the therapeutic intervention during the disease condition would surely impact the lives of elderly population.

502957 Principles for Biosensing Based on Liquid Crystals

Abstract:

Liquid crystals are fluid yet ordered phases that are well-known to assume orientations near interfaces that reflect the molecular structure and organization of adsorbates. These general characteristics make liquid crystalline phases promising candidates for the design of materials that can amplify the presence of biological molecules and their assemblies into optical outputs for biological sensing. This talk will address several key challenges that we are exploring to realize this potential. A first key challenge relates to the design of amphiphilic molecules that can couple biomolecular interactions to ordering transitions in liquid crystals. This presentation will describe a general approach that revolves around the use of amphiphilic copolymers that integrate monomers with side chains that either encode specific biological interactions or couple the organizations of the polymers to the ordering of liquid crystals. A second key challenge addresses the influence of size and shape of liquid crystalline domains on responses to biomolecular interactions. Specifically, I will describe how the elastic properties of liquid crystals can be manipulated by geometry to create exquisitely sensitive sensors of targeted biological molecules (proteins and lipids). Finally, the challenge of optical transduction will be described. In particular, the use of low-cost, flow focusing and light scattering methods will be shown to permit rapid, quantitative characterization of the internal configurations of micrometer-scale liquid crystalline droplets.

502917 In Search of an In Vivo Biopsy: Studies in Stimulus-Responsive Colloids for Biosensing

Abstract:

Early disease detection and diagnosis requires sensing of minute quantities of biomolecules in drawn fluids or image localized abnormalities in the body. Improvement upon existing technologies will require development of agents that can change their structure in response to the presence of an elevated biomarker with sufficient sensitivity, ignore the many other biomolecules present, and signal a positive result. This talk will address efforts in the Goodwin Lab to develop such technologies based on interactions between colloids and analytes that result in changes in detectable signal by many orders of magnitude. We primarily utilize the nucleic acid-driven formation of colloidal dimers for detection of biomolecules, utilizing the vast difference in signal generation between dimers and unimers. In one case, we developed mutually-reactive reagents that produce a fluorescent signal after aggregation and content mixing. In another case, the alignment of fluorocarbon droplets in an acoustic field allowed vaporization and detection by clinically relevant ultrasound. Each of these approaches utilizes mechanisms for in situ biomolecular detection, facilitating the development of in situ biosensors without washing steps or microfluidics.

502827 Printed Electronic Biosensors for Protein Detection

Abstract:

Floating gate transistor (FGT) sensors offer the potential for label-free detection of biomolecules with high sensitivity. In the typical ChemFET sensing strategy, the transistor is immersed in the sample solution. Binding of an analyte to the gate electrode or the semiconductor turns the transistor ON and produces the signal. This approach imposes substantial constraints on the materials used for the transistor, since one needs to balance the electronic properties of the semiconductor with its stability in the (aqueous) sensing environment.

To solve this problem with ChemFETs, we have developed a side-gated FGT sensing strategy that decouples the electronics from the sensing environment. In our approach, the semiconductor channel is capacitively coupled to a floating gate, the latter being functionalized with an antibody or aptamer for the desired analyte, and only the floating gate is available for binding. The potential of the floating gate is controlled by the main gate electrode, which is also in contact with an electrolyte solution. Binding of the target molecule to the antibody or aptamer film on the floating gate can turn the transistor ON. To enhance sensitivity and speed of analysis, we have augmented this basic architecture with an ion-gel dielectric, which provides large charge separation at relatively modest applied potentials and allows us to fabricate the electronics using aerosol jet printing, and a microfluidic system that permits rapid mass transfer to the sensor surface and prevents contact between the main gate and the sample solution.

In the first part of this presentation, I will present a series of fundamental experiments that we have used to uncover the details of the sensing mechanism and design rules to enhance sensitivity. In the second part, I will present results for the detection of increasingly challenging analytes, moving from DNA hybridization to ricin to gluten, in challenging matrices such as liquid foods and strongly reducing media. In particular, these applications demonstrate the ability of this sensing strategy to achieve quantitative detection and limits-of-detection at application relevant levels (e.g., â€œgluten freeâ€) without the need for any pre-concentration steps or sample cleanup.

References:

Scott P. White, Srinand Sreevatsan, C. Daniel Frisbie and Kevin D. Dorfman, Rapid, selective, label-free aptameric capture and detection of ricin using a printed floating gate transistor. ACS Sensors 1, 1213-1216 (2016).

Scott P. White, Kevin D. Dorfman and C. Daniel Frisbie, Operating and sensing mechanism of electrolyte-gated transistors with floating gates: Building a platform for amplified biodetection. Journal of Physical Chemistry C 120, 108-117 (2016).

Scott P. White, Kevin D. Dorfman and C. Daniel Frisbie, Label-free DNA sensing platform with low-voltage electrolyte-gated transistors. Analytical Chemistry 87, 1861-1866 (2015).

502828 Nanostructures for Biosensing

Abstract:

Recent technological advances have enabled high-throughput fabrication of plasmonic nanogap structures with critical dimensions approaching 1 nm. Extreme confinement of optical energy in such single-nanometer gaps can lead to a series of new applications in biosensing and spectroscopy. This presentation will outline approaches to design and fabricate nanogap plasmonic sensors with precisely tuned optical resonances in the visible, near-infrared, or mid-infrared regime. Among various options, we use atomic layer deposition (ALD) to create a dense array of ultra-narrow slits and rings in metals, which is made possible by the Angstrom-scale thickness resolution of ALD. Our scheme â€" atomic layer lithography - has enabled rapid wafer-scale production of various nanostructures that are nontrivial to produce using any other methods. For example, now it is a routine task to produce a dense wafer-scale array of 1-nm-wide metal slit that is as long as ~1 cm in length. Metasurfaces made with such nanoslits and rings that are resonant in the visible, infrared, and even terahertz regime have been produced over a 4-inch wafer.

Since the nanogap size in our scheme can be precisely controlled with ALD to match the size of target biomolecules, the resulting nanogap structures can be used for trapping and detection of single molecules and nanoparticles. High-throughput production of these nanostructures allows researchers to perform such challenging experiments in a more reproducible manner. We will present experimental results on how to combine optical and electronic trapping techniques with these nanoplasmonic sensors to accomplish single-molecule trapping and detection.

492677 Point-of-Care Determining Small Molecule Drug with Multi Hydrogen Bonding Manipulated Single-Molecule Recognization

Abstract:

An electrochemical sensor based on functionalized graphene (Arg-G) nanocomposite was fabricated for sensitive detection of acetaminophen (AP) via mutilple hydrogen bonding platform. The electrochemical behaviors of acetaminophen on Arg-G composite film modified glassy carbon electrode were investigated by cyclic voltammetry, quartz crystal microbalance and differential pulse voltammetry. The interaction between Arg and AP also were investigated by Fourier Transform Infrared Spectroscopy, UV photometer and fluorescence spectra. The results indicated that Arg could form hydrogen bond with AP, which could enhanced the electrochemical response of acetaminophen. This fabricated sensor displayed excellent analytical performance for acetaminophen detection over a range from 0.1 to 3000 Î¼mol $L^{â"1}$ with a detection limit of 0.05 Î¼mol L^{-1} (S/N=3). The sensor also showed good performance for AP detection in serum. Moreover, the proposed electrochemical sensor also exhibited good reproducibility and stability, and has been used to detect acetaminophen in tablets and the serum.

492839 Analysis of Multiplexed Nanosensor Arrays Based on Nir Fluorescent Single Walled Carbon Nanotubes

Abstract:

The high-throughput, label-free detection of biomolecules remains an important challenge in analytical chemistry with the potential of nanosensor technology to significantly increase the ability to multiplex such assays. Herein, we analyze the variance and binding kinetics of a nanosensor array based on fluorescent single wall carbon nanotubes. The nanotubes are non-covalently modified with Cu^{2+} chelator - nitrilotriacetic acid group â€" to immobilize immunobinding proteins, such as Protein A, G and L as molecular recognition sites. The Cu^{2+} ions also serve as a proximity quencher that when the analyte antibodies bind with the immunobinding proteins, the resulted distance change between the Cu^{2+} ions and the nanotube surface modulates the near IR emission of nanotubes and gives a direct readout. Based on this mechanism, we develop an optical microarray composed of sensor spots about 200 um at a density of ~300 spots/cm^2, which is capable of real-time and quantitative detection of Mouse IgM, Human IgD and Rat IgG2a, in the concentration range between 125 â€" 500 ug/mL. Moreover, the binding forward reaction rates can be extracted from the initial binding response, which are significantly different among the recognition sites. In order to perform multiplexed detections, the sensor array is modified with Protein G and Protein L in alternative columns, showing distinct binding affinities with the same analyte, demonstrating the capacity to perform high throughput detection with the optical microarray. In addition to the platform development, we also investigate the spot-spot variance, which is shown to result from the inhomogeneous distribution of nanosensors in the spot during the printing and modification process.

496970 A Biomimetic Tongue By Photoluminescent Metal–Organic Frameworks

Abstract:

The taste sensing capabilities of a "biomimetic tongue" based on the photoluminescence (PL) responses of metal-organic frameworks (MOFs), [In(OH)(bdc)]*n* (bdc=1,4-benzenedicarboxylate), [Tb(btc)]*n* (MOF-76, btc=benzene-1,3,5-tricarboxylate), and $[Ca_3(btc)_2(DMF)_2(H_2O)_2]\cdot 3H_2O$ are proven on aqueous solutions of five basic tastants: sucrose (sweet), caffeine (bitter), citric acid (sour), sodium chloride (salty) and monosodium glutamate (umami). For [In(OH)(bdc)]*n*, the tastant interacts stereochemically with poly(acrylic acid) (PAA) and alters its conformations. The frequency and magnitude of chelation between COO^- pendant groups of PAA and In^{3+}nodes of [In(OH)(bdc)]*n* framework influence the corresponding PL reponses. For MOF-76, the tastant interacts with incorporated water in MOF-76 through hydrogen bonding. The limitation of O–H bond stretching of water results in the enhancement of the PL intensity. For $[Ca_3(BTC)_2(DMF)_2(H_2O)_2]\cdot 3H_2O$, it is added as a third MOF component to increase the precision on taste discrimination. The significance of MOF-based "biomimetic tongue" includes: (1) PAA on [In(OH)(bdc)]*n* mimics the taste receptor cells (TRCs) for their structural flexibility, (2) the Weber-Fechner law of human sensing that sensation is proportional to the logarithm of the stimulus intensity is observed between the PL emission response of MOF-76 and the concentration of tastant, (3) the strength of taste is quantified by the τ scale and the PL emission intensity of MOF-76, which are dependent on the logarithmic tastant concentration, (4) the tastant is identified by the shape of the 3D principal component analysis contour map (i.e., pattern recognition method), and (5) the fabrication of [In(OH)(bdc)]*n*/PAA film by brushing is illustrated.

499490 Outstanding Surface Plasmon Resonance Sensitivity and Figure of Merit Enabled By Periodic Gratings Templated from Optical Discs

Abstract:

Surface plasmon resonance (SPR) is a powerful and widely exploited technology for ultrasensitive chemical and biological sensing. Unfortunately, it is generally challenging to simultaneously achieve very high SPR sensitivity and figure of merit (FOM) by using plasmonic nanostructures that can be scalably and economically produced. Here we report a simple and scalable soft-lithography-based templating technology for fabricating plasmonic gratings by using optical discs (DVDs and CDs) as structural template. The resulting Au-coated gratings simultaneously exhibit very high surface plasmon resonance (SPR) sensitivity (up to ~940 nm per refractive index unit, nm/RIU) and figure of merit (FOM, up to ~110). Systematic finite-difference time-domain (FDTD) simulations reveal that the SPR properties (e.g., dip position, amplitude, sensitivity and FOM) of the templated gratings are sensitive functions of the geometries of the track-pitch structures of the optical discs. The numerical simulations agree well with the experimental results. The outstanding SPR performance combined with the wide varieties of functional materials that can be templated (e.g., titania with high thermal stability), as well as the scalability and cost benefits of this bottom-up nanofabrication approach, could make the templated plasmonic gratings a promising SPR platform for many important chemical and biological sensing applications, such as real-time and label-free sensing of biomolecular adsorption, and in-situ probing heterogeneous catalytic reactions under realistic conditions.

489935 Development of Highly Sensitive Pico-Calorimetric Sensors Based on Thermoelectric Effect

Abstract:

Microcalorimeters have great potential as biosensors due to their ability to make measurements on very small samples and their short thermal equilibration times. However, they are not yet sensitive enough to analyze nano/micro-scale samples or individual cells, and liquid sample handling remains a critical issue. This work aims to break through these current limitations by measuring pico-Watt power levels quantitatively through use of micromachined sensors and by controlling the wetting and evaporation of liquids. Individual sensors consist of a combination of thermistors and thermopiles on a thin silicon nitride membrane that allow direct differential measurements between a sample and a reference. To maximize sensitivity, sensor geometry and materials were optimized based on the temperature field that develops around a typical sample and on calculations of the corresponding noise equivalent power and minimum detectable power (P_{min}). Specifically, thermopiles that consist of Constantan and Nichrome couples are capable of sensing below nW level power over a broad range of thermopile lengths, 10 - 10^3 μm. Further, thermopile responses at the given power were characterized as a function of thermopile dimensions and liquid volumes. This highly sensitive thermopile-based calorimeter offers particular promise for quantitative measurement of cellular bioenergetics and systems that are limited in analytic volume.

A Micro-Fabricated Electrochemical Gas Sensor Tailored for the Detection of Methane

P.A. Gross, E. Sadeghipour, T.F. Jaramillo and B.L. Pruitt
Stanford University, USA

Abstract

We present the fabrication, characterization, and testing of an electrochemical VOCs sensor operating in gaseous condition at room temperature. It is designed to be microfabricated and to be operated using Cyclic Voltammetry (CVs). It is composed of a Working Electrode (WE), a Counter (CE) Electrode, a Ag Reference Electrode (RE), and a *Nafion* solid-state electrolyte. This polymer conducts protons (H^+) from the WE to the CE, which are generated from redox reactions of ambient gases. We have shown that our sensor is not only capable of detecting and quantifying the concentration of methane, but also it can differentiate the signal of methane form that of carbon monoxide (CO). These results have been confirmed by exposing the sensor to two dilution levels (50% and 10% dilution in N_2) of methane, as well as pure CO.

Introduction

Continuous atmospheric monitoring is particularly challenging due to the lack of an inexpensive, low power sensor that can reliably detect multiple molecules in ambient air. Sensors using metal-oxides are the closest technologies capable of enabling a large distributed gas sensor network. These sensors measure the change in resistance between two electrodes as molecules are adsorbed on the surface of the medium in between them[1,2]. However, some major drawbacks are limiting the deployment of this kind of sensor. For example, these sensors must be operated at fairly high temperatures (300 – 500 °C), and there is no ability for the identification and quantification of detected molecules. Solutions to circumvent these issues consist of assembling multiple such sensors in arrays, where each sensor utilizes a metal oxide most sensitive to of them having a sensing layer more sensitive to certain molecules. Nevertheless, the small form-factor of these sensors and their compatibility with mass fabrication techniques makes them very attractive, and such advantages should be conserved when developing other kinds of gas sensors.

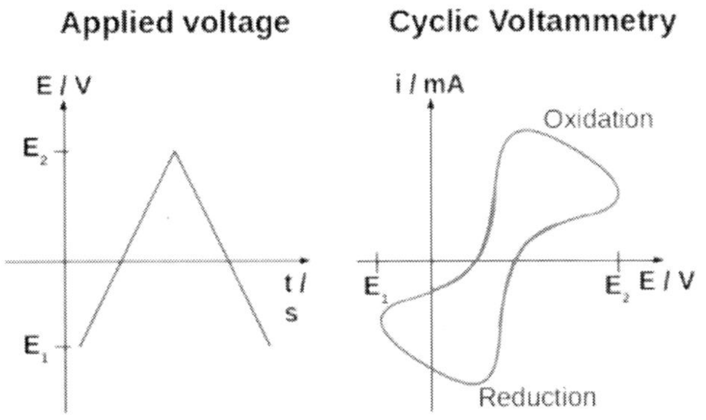

Figure 1: Cyclic Voltammetry technique. The voltage is applied between the Working (WE) and the Counter (CaE) electrodes (left) and the current is plotted against the applied voltage to obtain the CV curve (right).

From these observations, we have developed a gas sensor chip that keeps these advantages but tackles the drawbacks by using a different sensing mechanism. Rather that measuring a simple change in resistance, which provides little information, we use the characteristic redox potential of molecules to detect, identify, and quantify them. The redox potential corresponds to the amount of energy, in the form of applied voltage, that is necessary to trigger the oxidation or the reduction of a molecule. Each Volatile Organic Compound (VOC) has a different redox potential that can be used to detect it. The molecules detected in the present paper are methane (CH_4) and carbon monoxide (CO), with redox potentials of 0.17 V vs the Standard Hydrogen Electrode (SHE) and 0.52 V vs SHE, respectively.

In order to use the redox potential for detection purposes, it is necessary to use an electrochemical technique called cyclic voltammetry (CV). This technique consists of sweeping a voltage between two electrodes and measuring the current that is generated. When the voltage corresponds to the redox potential of a molecule adsorbed on the electrodes, a current is measured. When the current is plotted against the applied voltage, the I-V curve exhibits current peaks at particular voltages that correspond to the molecules present on the surface of the electrode. A scheme of the CV technique is presented in **Fig. 1**.

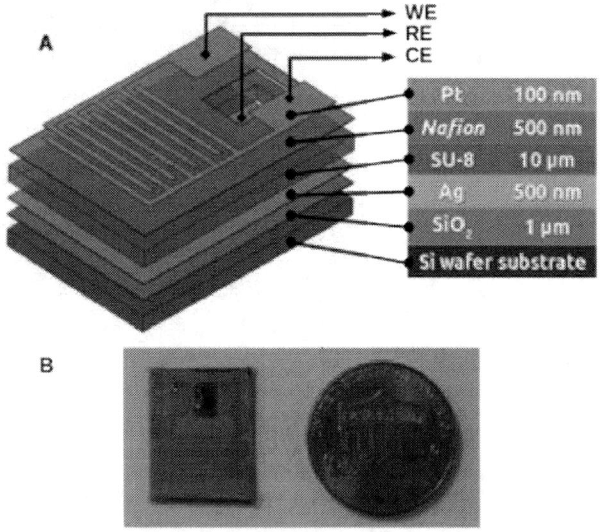

Figure 2: (A) The sensor is composed of a reference Ag layer, an SU-8 adhesion layer, a *Nafion* layer and two interdigitated Pt electrodes all on a Si/SiO_2 substrate. The Ag layer is made accessible from the top by cutting through the two polymers. The thicknesses indicated for each layer are the targeted ones. (B) The sensor has a size comparable to a US dollar penny.

Although CV is a well-known technique in the liquid phase, its application in the gas phase is less trivial. We have identified the four key components necessary to enable such measurements, three electrodes, a *Working* (WE), a *Counter* (CE) a *Reference* (RE), and an electrolyte to conduct ionic products from the WE to CE. For application in the gas phase, the electrolyte has to be solid-state, and we have decided to work with *Nafion. Nafion* is a fluorinated polymer, which conducts protons (H^+), and is traditionally used in fuel cells. The presence of the RE is necessary to be able to correlate the position of the observed current peaks on the CVs with the redox potential values which are tabulated against the SHE.

Design and Fabrication

The simple design and fabrication of our sensor chip, and its compatibility with standard micro-fabrication techniques makes it a good fit

Table 1: The measured thicknesses of the layers composing the sensor are in good accordance with the projected once. The *Nafion* layer shows higher thickness variation because it is drop-casted.

Layer	**Thicknesses (nm)**	
	Target	**Measured**
Oxide – SiO_2	1,000	920
RE – Ag	500	510
Adhesion – SU-8	10,000	9,600
Electrolyte – *Nafion*	500	600 – 800
WE and CE – Pt	100	110 – 120

for our vision of a large-scale distributed gas sensor network.

Fabrication process: A 100 mm diameter, 525 µm thick Si wafer is cleaned using a succession of baths: first a bath of 9:1 H_2SO_4:H_2O_2 (piranha), second a bath of 5:1:1 H_2O:H_2O_2:NH_4OH, third a bath of 50:1 HF, and finally a bath of 5:1:1 H_2O:H_2O_2:HCl. A 1 µm thick Si oxide (SiO_2) layer is then grown by wet oxidation at 1100 °C for 2.25 hrs. A 10 µm thick SU-8 *2010* layer is spin coated on top of the oxide at 1500 rpm for 5 s and then 3000 rpm for 15 s. The layer is exposed to a 119 mJ/cm^2 dose of 365 nm ultraviolet light, baked at 85 °C for 2 min to ensure hardening and drying of the SU-8 and then developed. This layer is subjected to 30 s of O_2 plasma prior to the deposition of the *Nafion* layer. *Nafion* is deposited by two consecutive drop castings of 5 mL of a *Nafion® D1021* water dispersion directly on the substrate sitting on a hot plate heated to 100 °C, ensuring optimal coverage of the surface. The solvent is evaporated between the castings, and for 10 min at the end. This deposition process showed better results in layer homogeneity and thickness than spin coating. To complete the layer stack showed on **Fig. 2A** a 100 nm thick Pt layer is deposited by e-beam evaporation, at a 1 Å/s rate, through a shadow mask with the geometry of interdigitated electrodes. Once completed, the sensors are diced out of the wafer using a diamond tip and an electrical access to the Ag layer is cut out through the *Nafion* and SU-8 membranes using a razor blade. The sensors are then characterized, activated in a liquid environment (standard for *Nafion*) and tested for the detection of various gases.

Characterization: All topological characterizations were made using a profilometer and the results are presented in **Table 1**. The measured thicknesses presented in the table are averaged over the entire wafer and are close to the target thicknesses. The *Nafion* layer shows some disparity across the

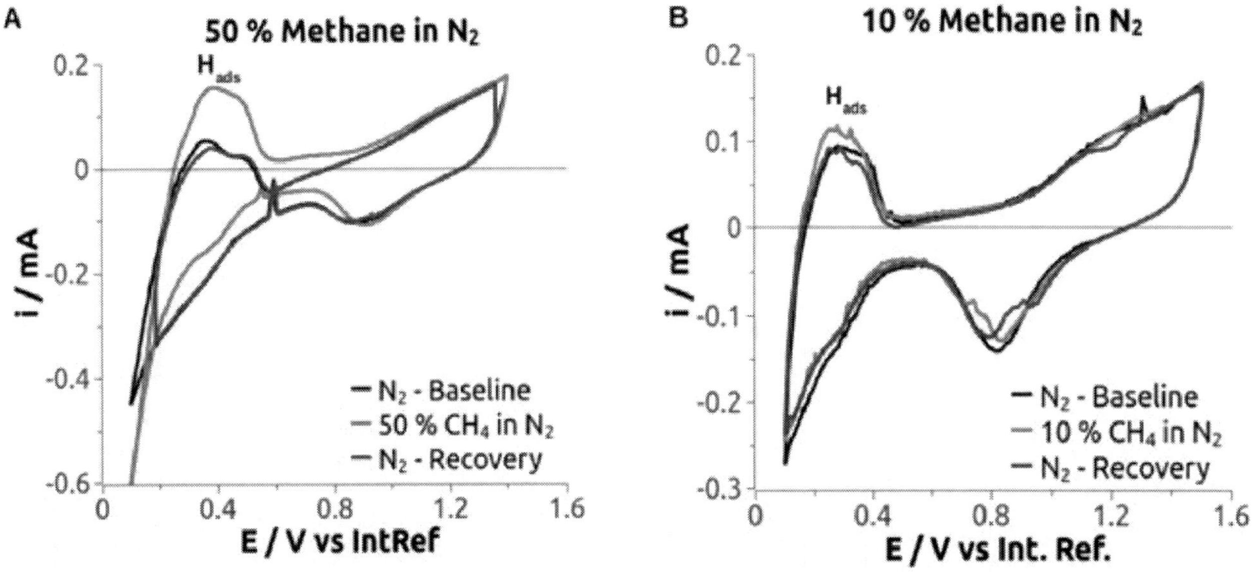

Figure 3: When methane is present in the gas flow, the H_{ads} region increases and reaches a maximum. This increase is higher for 50 % (A) of methane in the gas flow compared to 10 % (B). The position of this increased signal is located at 0.38 V vs IntRef. The curve gets back to its original shape when methane is turned off. Sweep rate is 100 mV/s.

wafer, but is thicker and more homogeneous than previously published when it was deposited using spin coating.

Results and Discussion

All experiments performed with the sensor are carried out using a test setup composed of a glass cell in which the sensor is placed, and a potentiostat connected to the sensor. Mixtures of gases flow through the glass cell using Micro-Flow Controllers (MFCs) and a mixing chamber. All gas flows are expressed in *Standard Cubic Centimeter* (sccm). The proportion of each gas is expressed in percentage of the total gas flow that exits the mixing chamber. The total gas flow is maintained at a constant 20 sccm. The three electrodes (WE, CE, RE)of the sensor are connected to a *Biologic®* SP-300 potentiostat that performs cyclic voltammetry.

<u>Testing in gas</u>: Once the sensor has been activated in liquid, the surface of the sensor is dried out with an air stream and is used for gas detection experiments. Typical experiments are carried out by first acquiring a baseline curve with only N_2 flowing through the cell, then the N_2 is adjusted proportionally as Methane is added to the flow to keep the total flow constant. Two detection experiments were performed with Methane: (i) 50 % Methane in N_2 which corresponds to 10 sccm of Methane plus 10 sccm of N_2 and (ii) 10 % of Methane in N_2 which corresponds to 2 sccm of Methane plus 18 sccm of N_2. Finally, the Methane is turned off and N_2 flows alone in the cell to follow the recovery of the sensor as the curve returns back to the baseline. Detection results are presented in **Fig. 3A** and 3B, which show raw CVs acquired during the testing experiment described earlier, with 50 % and 10 % of methane mixed in N_2. The main observation

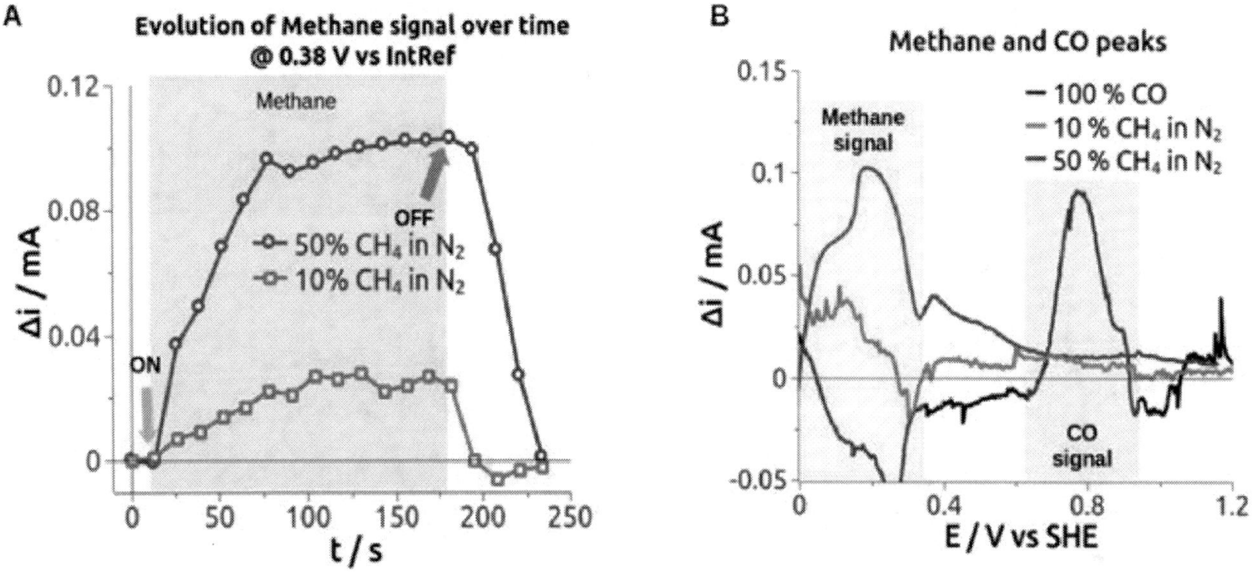

Figure 4: (A) Once the baseline subtracted the signals with different concentration of methane can be compared. The amplitude of the signal is proportional to the concentration of methane in the flow showing the quantification capabilities of the sensor. (B) The position of the methane peaks observed on CVs can be clearly identified and are well separated from a signal obtained under CO flow in similar conditions showing the speciation capabilities of the sensor.

that can be made is that the increase in current from the H_{ads} region of the Pt curve located between 0.1 and 0.6 V vs IntRef. The position of this current increase is difficult to attribute precisely to direct oxidation of methane into CO_2. Although the standard redox potential for the oxidation of methane into CO_2 is indeed in that region (0.17 V vs SHE[3]) it requires some over-potential to occur. Methane is known to be the least reactive hydrocarbon at room temperature[4], which is why its detection is so difficult. Thus, in the present conditions, observing direct electro-oxidation of methane into CO_2 is unlikely. However, observing a stable, reproducible, and proportional signal, even from indirect electrochemical processes, is very encouraging for the detection of this molecule especially with a sensor with such a simple design and fabrication process and at room temperature.

Signal treatment: The raw CVs are further transformed using mathematical operations, which consist of correcting the signals by subtracting the baseline and applying the reference electrode shift determined during the activation experiment. Two types of data sets are presented on **Fig. 4. Fig. 4A** shows the evolution of the methane signal as a function of time, and **Fig. 4B** shows the position of the methane signal compared to another VOC, carbon monoxide (CO).

On **Fig. 4A**, the value of the current of the CVs from **Figs. 3** at 0.38 V vs IntRef is plotted against time. The value of time is known from the sweep rate of the CVs which is 100 mV/s. On these curves, one can see that the signal increases as soon as methane is flown in the cell, and continues to increase until it reaches a steady state. The sensor shows a good response time, of *ca.* 13 s (the time needed to complete a full voltage cycle). The fact that the curve reaches a steady state shows that the Pt surfaces become saturated and cannot adsorb more methane. These observations are consistent when the concentration of methane in the flow is decreased from 50 to 10 %. Moreover, the change in the signal is proportional to the concentration of methane as it is about five times higher with 50 % of methane than with 10 %. This proportionality shows the possibility of using the electrical current value to quantify the concentration of methane. When methane is removed from the gas flow, it only takes a few cycles to go back to the baseline value, showing the good recovery time (in the order of tens of seconds to a minute) of the sensor. The signal going back to its original value shows also that the surface does not get poisoned. This is also an advantage of the cycling, as by applying higher voltages, all molecules eventually get oxidized and desorb.

The curves shown in **Fig. 4B** are obtained by baseline correction, as well as the application of the reference electrode shift measured during the activation experiment performed in liquid. First, the blue curve (50 % methane is N_2) shows a clear peak centered at 0.2 V vs SHE, which corresponds to the signal observed on **Fig. 3A**. This peak is the main detection signal coming from the presence of methane in the gas flow. When the concentration of methane is decreased to 10 %, one can see that the detection signal becomes weaker and the signal to noise ratio is important. Nevertheless, the maximum of the red curve is in the same region as the blue one suggesting that similar electrochemical processes are happening at lower concentrations.

The methane signals are compared to a signal obtained with CO (black curve) in similar conditions and processed using the same mathematical treatments. This experiment is presented in greater detail in our previous publication for the detection of CO with an earlier design of the sensor[5]. The peak observed on the black curve corresponds to the CO oxidation into CO_2, and is located at 0.8 V vs SHE, and is clearly separated from the signals obtained with methane. This result shows that cyclic voltammetry can differentiate molecules using their respective redox potential by exhibiting curves with different peaks for different chemical species.

Conclusion

We have presented a gas sensor design based on electrochemical measurement, that performs cyclic voltammetry. The sensor has been used to prove the capability of this sensing principle to detect methane dissolved in N_2. The results showed clear and reproducible oxidation signals that are attributed to the presence of methane in the gas flow. The position of this signal for methane is compared to CO, and is found to be clearly separated from it, showing the speciation capabilities of the sensor. In addition, our experiments have shown that it is possible to use the current value to quantify the detected molecule in the gas flow. We have successfully demonstrated the advantages of our micro-fabricated electrochemical gas sensor and its sensing principle and that it can now be applied to more advanced designs to determine crucial specifications such as detection limits, response time, recovery time, as well as lifetime. We can also now move towards building and testing field-deployable devices.

References

1. Liu, X. *et al. Sensors* **12,** p9635 (2012).
2. Huber, F., *et al. Sens. Act. B Chem.* **239,** p358, (2017).
3. Eng, D. *et al. J. Catal.* **130,** p306 (1991).
4. Otagawa, T. *et al. J. Electrochem. Soc.* **132,** p2951 (1985).
5. Gross, P.-A. *et al. Micro Nano Lett.* **11**, p798 (2016).

497639 Highly Selective, Flame-Made Sensors for Breath Analysis

Abstract:

Highly selective, flame-made sensors for breath analysis

Andreas T. Güntner and **Sotiris E. Pratsinis**

Particle Technology Laboratory, Institute of Process Engineering,

Department of Mechanical & Process Engineering,

ETH Zurich, CH-8092 Switzerland

Flame aerosol technology dominates the manufacture of nanostructured materials at tons/h, albeit rather simple ones (carbon blacks, fumed silica, alumina, titania etc.). Recent advances in combustion and aerosol science with multi-scale process design bring this technology to synthesis of far more sophisticated materials (e.g.nanosilver, subnanoclustered catalysts & biomedical ferrofluids) and even devices[1], such as portable gas sensors for breath analysis that are the focus of this presentation.

Taking advantage of the capacity of combustion for stable synthesis of metastable phases and solid solutions, flame aerosol deposition and *in-situ* annealing of gas sensor films is presented as it had led, first, to optimally-doped SnO_2 for ethanol and CO sensing and, most notably to epsilon-WO_3 for selective detection of acetone (a diabetes type-1 tracer in the breath) at the ppb level & 90% relative humidity (RH). In tandem with PTR-MS, such a portable breath sensor was used for online and offline testing of humans benchmarked with standard glucose tests (finger pricking) [2]. This led to an industrial prototype for clinical testing[3] while MoO_3- and ZnO-based sensors are developed for kidney disease[4] and cholesterol[5]monitoring, respectively.

Emphasis now is placed in development of sensor arrays (E-nose) with focus on formaldehyde (FA) as it is a potential breath marker for lung cancer and a tracer for indoor air quality monitoring[6]. Typical FA concentrations are below 100 ppb posing a sensitivity and selectivity challenge. Here, I present a highly sensitive, selective and compact electronic nose (E-nose) for on-line quantification of FA in realistic gas mixtures. This E-nose consists of four nanostructured and highly porous Pt-, Si-, Pd- and Ti-doped SnO_2 sensing films. The constituent sensors offer stable responses and detection of FA down to 5 ppb (signal-to-noise ratio > 30) at breath-realistic 90% RH. Each dopant induces different analyte selectivity enabling *selective* detection of FA in gas mixtures by multivariate linear regression. In simulated breath (FA with higher acetone, $NH_3$and ethanol concentrations), FA is detected with an average error ≤ 9 ppb overcoming selectivity issues of single sensors. This device can facilitate easy screening of lung cancer patients and

monitoring of indoor FA concentrations.

1. Aerosol-based Technologies in Nanoscale Manufacturing: from Functional Materials to Devices through Core Chemical Engineering, *AIChE J.*, **56**, 3028-3035 (2010).

2. M. Righettoni, A. Schmidt, A. Amman, S.E. Pratsinis, "Correlations between blood glucose & breath components from portable gas sensors and PTR-TOF-MS", *J. Breath Res.*, **7**, 037110 (2013).

3. M. Righettoni, A. Ragnoni, A.T. Güntner, C. Loccioni, S.E. Pratsinis, T.H. Risby, "Monitoring breath markers under controlled conditions", *J Breath Res.*, **9**, 047101 (2015).

4. A.T. Güntner, M. Righettoni, S.E. Pratsinis, "Selective sensing of NH_3 by Si-doped α-MoO_3 for breath analysis", *Sensors and Actuators B: Chemical*, **223**, 266–273 (2016).

5. A.T. Güntner, N. Pinau, D. Chie, F. Krumeich, S.E. Pratsinis, "Selective sensing of isoprene by Ti-doped ZnO for breath diagnostics", *J. Mater. Chem. B*, **4**, 5358 - 5366 (2016).

6. A.T. Güntner, V. Koren, K. Chikkadi, M. Righettoni, S.E. Pratsinis, "E-nose sensing of low-ppb formaldehyde in gas mixtures at high relative humidity for breath screening of lung cancer?", *ACS Sensors*, **1**, 528-535 (2016).

498572 A Drinking Water Sensor for Lead and Other Heavy Metals

Abstract:

Leakage of lead and other heavy metals into drinking water is a significant health risk to humans and a major concern in the United States. Heavy metal leakage often occurs without the awareness of the users, as lead can be colorless and odorless. Early detection of lead is important to prevent long-term exposure but is difficult to achieve using current technology. Because water is contaminated inside the structure of a house, end-point detection by home-monitoring is crucial for lead leakage detection.

We have developed sensors consisting of platinum electrodes for the detection of heavy metal that have a long lifetime, cost around 10 cents per unit, and can be easily inserted in pipes for continuous monitoring. The sensors operate by measuring the impedance change between the electrodes. Two designs were created: A two-electrode sensor that can identify the existence of a variety of heavy metals, and a four-electrode sensor that can distinguish lead from other heavy metals. No false-positive response was generated when the sensors were placed in simulated and actual tap water contaminated by heavy metals. Lead detection on the four-electrode sensor was not affected by the presence of other common ions in tap water.

499205 Real-Time Underwater Detection of Trace Organic Analytes

Abstract:

Underwater detection of trace analytes requires a sensor system with superb sensitivity because of the low solubility of most organic compounds in water. Furthermore, analytes rapidly mix in the water, and prevailing currents make the concentration highly localized and non-uniform. In addition, chemical and biological interferents in the seawater can introduce bias and complicate analysis. To address this, a microfluidic, surfaced enhanced Raman scattering (SERS) device was developed to detect analytes of interest at low part per billion (ppb) concentrations in seawater. The multi-analyte sensor is compact, low-power, and ruggedize for a marine environment aboard an autonomous underwater vehicle (AUV). Lab-scale demonstrations were conducted to validate instrument performance and determine practical limits of detection in a seawater matrix. Sensor development, analytes tested, and operating parameters (e.g., limits of detection, response time) will be presented.

500217 Application of Hydrophobic-Polymer Coated TiO_2 Nanotube Electrochemical Sensors in Humid Environments

Abstract:

In this work, the effect of a hydrophobic, gas permeable polymer coatings on a vapor-based electrochemical sensor is investigated. This technology will allow volatile organic compounds (VOCs) to be sensed more accurately in humid environments, such as those found in human breath, agricultural settings, and industrial processes. In this case, cobalt-functionalized titanium dioxide nanotube array (Co-TNA) sensors were coated with a thin layer of cross-linked polydimethylsiloxane (PDMS). These sensors are traditionally used uncoated in the rapid detection of mycobacterium tuberculosis in human breath. This is achieved by detecting several different VOC biomarkers produced by the mycobacterium, the most important of which is methyl nicotinate. The functionalized sensors are known to be sensitive to humidity causing issues with signal to noise ratio during operation. This is thought to be primarily due to water vapor condensing on the surface of the sensor, drawing cobalt ions away from nanotubes and making the signal less distinguishable than in dry air. A hydrophobic, gas permeable coating can be used to bypass this issue. PDMS is commercially available, its interactions are well known and it can easily be coated onto surfaces before it is fully crosslinked, making it an ideal candidate. The thickness and chemical transport properties of the polymer can be changed easily by diluting the PDMS in toluene before dip-coating and varying the amount of cross-linker added. Preliminary testing has shown that thinner coatings of PDMS allow the VOC (in this case methyl nicotinate) to reach the sensor surface more quickly than thicker coatings, as expected. For example, a dilution of 20 g of PDMS in 20 mL of toluene showed a response to the VOC after 20 min of exposure, while the same amount of PDMS in 40 mL of toluene responded after only 12.5 min. Methods for applying these thin coatings, transport properties of the VOC through the coating and the effect on sensor response will be presented. This work has applications in biomedical diagnostics, environmental/food safety, and water monitoring.

501144 Micro-Tensiometer -- a Sensor That Measures the Chemical Potential of Water

Abstract:

Climate change has caused extreme weather conditions, resulted in substantial variations in precipitation frequency, and presented particular challenges to plant life in both natural and agricultural contexts. It is important to study how the plants adapt to the water stress, not only to improve the water use efficiency in agriculture, but also to understand how the forests respond to climate change, and how, inversely, their responses could potentially impact climate change. In an era of high-technology, we are still lacking an accurate and robust tool to monitor the water stress of plants in an automated, cost-effective manner.

I would like to introduce a newly designed water stress sensor, a micro-tensiometer (µTM). This microelectromechanical system translates the chemical potential of water into mechanical stress – tension – within an internal liquid volume that, in turn, is measured with a strain gauge. The µTM sensor is as small as a pen, and can be easily installed into any water containing material of interest – soil, plants, food, cosmetics, medicines, and cements – to provide continuous measurements of the chemical potential.

In this talk, I will: 1) introduce the thermodynamics and transport phenomena of water motion in plants; 2) explain the plant-mimetic design principles of our mTM and its applications in measuring water stress in plants; 3) describe our continuous measurements of water stress in apple trees to provide unprecedented insights into the transient response of plants to varying environmental parameters; and 4) introduce a preliminary mathematical framework that accounts for both physical (e.g., poroelastic diffusion) and biological (e.g., active regulation of boundary layer resistance) to model the observed dynamics. I will conclude with perspectives on using the µTMs as part of the agricultural precise irrigation system and provide a feedback loop for controlled water-delivery, as well as studying the water relations of redwoods in response to the climate change.

AUTHOR INDEX